Some Power Electronics Applications Using Matlab Simpowersystem Toolbox

By
Dr. Hidaia Mamood Alassouli

1. Abstract

Matlab SimPowerSystems is a modern design tool that allows scientists and engineers to rapidly and easily build models that simulate power systems. Not only can you draw the circuit topology rapidly, but your analysis of the circuit can include interactions with mechanical, thermal, control, and other disciplines. The paper covers some case studies that provide detailed, realistic examples of how to use SimPowerSystems in power system analysis. The following types of studies are covered on the paper:

1. Thyristor-Based Static Var Compensator: Study the steady-state and dynamic performance of a static var compensator (SVC) on a transmission system.

2. Transient Stability of a Power System with SVC and PSS: Study of the application of static var compensator (SVC) and power system stabilizers (PSS) to improve transient stability and power oscillation damping of the system.

3. GTO-Based STATCOM: Study the steady-state and dynamic performance of a static synchronous compensator (STATCOM) on a transmission system.

4. Control of load flow using UPFC: Study the steady-state and dynamic performance of a unified power flow controller (UPFC).

5. Variable-frequency Induction Motor Drive: Study of a PWM inverter is used as a variable-voltage, variable-frequency source to drive an induction motor in variable-speed operation.

6. Chopper-Fed DC Motor Drive: Study of a DC motor drive with armature voltage controlled by a GTO thyristor chopper.

7. VSC-Based HVDC Link: Modeling of a forced-commutated voltage-sourced converter high-voltage direct current (VSC-HVDC) transmission link.

Index Terms--AC motor drives, DC motor drives, HVDC converters, Power system dynamic stability, Pulse width modulated power converters, Static VAR compensators, Thyristor converters.

2. Introduction

Electrical power systems are combinations of electrical circuits and electromechanical devices like motors and generators devices and sophisticated

control system concepts that tax traditional analysis tools and techniques. Further complicating the analyst's role is the fact that the system is often so nonlinear that the only way to understand it is through simulation. Land-based power generation from hydroelectric, steam, or other devices is not the only use of power systems. A common attribute of these systems is their use of power electronics and control systems to achieve their performance objectives. Matlab SimPowerSystems[1] is a modern design tool that allows scientists and engineers to rapidly and easily build models that simulate power systems. It uses the Simulink environment, allowing you to build a model using simple click and drag procedures. SimPowerSystems toolbox is an efficient tool for modeling and simulating power system with FACTS devices and power converters. SimPowerSystems also is an efficient tool for modeling the dynamics of power system including all types of machines and controllers, i.e. modeling synchronous machines with governor, exciter and AVR, power system stabilizer, asynchronous machines, High Voltage DC Links (HVDC), DC motors and various speed controllers for DC and AC drives using power electronics converters. Reference [3] is one of the best references in studying multimachine system dynamics and stability. Reference [4] includes analysis of the properties of semiconductor devices and their applications as controlled rectifiers, inverters, cycloconverters, choppers, HVDC and solid state control of DC and AC motor. Reference [2], [5] include an overview of the principles of work of most FACTS devices.

The paper covers some case studies that provide detailed, realistic examples of how to use SimPowerSystems in modeling power system dynamics in various types of application that use power electronics converters. The following case studies are simulated on the paper:
1- Thyristor-Based Static Var Compensator.
2. Transient Stability of a Power System with SVC and PSS.
3. GTO-Based STATCOM.
4. Control of load flow using UPFC.
5- Control of AC motor.
6- Control of DC motor.
7- VSC-Based HVDC Link.

3. THYRISTOR-BASED STATIC VAR COMPENSATOR

The SVC system shown in Fig.1, consists of a 735 kV/16 kV, 333 MVA coupling transformer, one 109 Mvar TCR bank and three 94 Mvar TSC banks (TSC1 TSC2 TSC3) connected on the secondary side of the transformer.

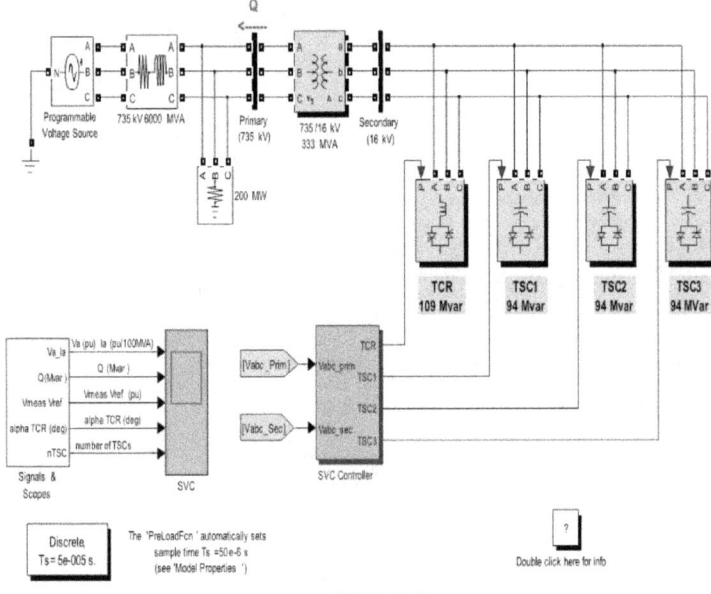

Fig. 1. SPS Model of the 300 Mvar on a 735 kV Power System

Switching the TSCs in and out allows a discrete variation of the secondary reactive power from zero to 282 Mvar capacitive (at 16 kV) by steps of 94 Mvar, whereas phase control of the TCR allows a continuous variation from zero to 109 Mvar inductive. Taking into account the leakage reactance of the transformer (0.15 pu), the SVC equivalent susceptance seen from the primary side can be varied continuously from -1.04 pu/100 MVA (fully inductive) to +3.23 pu/100 Mvar (fully capacitive). The SVC Controller monitors the primary voltage and sends appropriate pulses to the 24 thyristors (6 thyristors per three-phase bank) to obtain the susceptance required by the voltage regulator. The SVC control system is shown in Fig. 2, consists of the following four main modules:

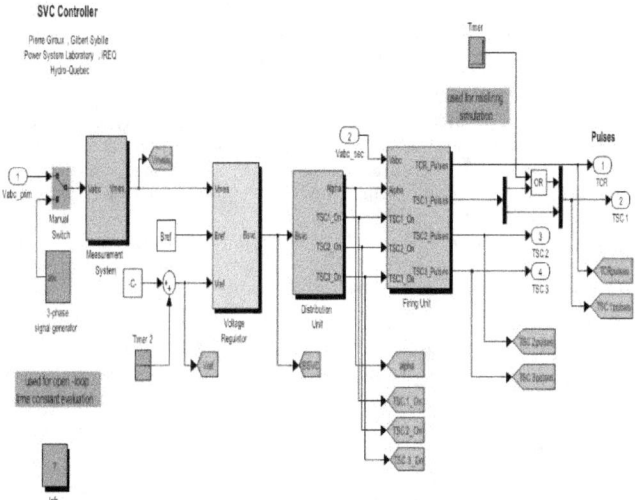

Fig. 2. SVC Controller

1) Measurement system measures the positive-sequence primary voltage. This system uses discrete Fourier computation technique to evaluate fundamental voltage over a one-cycle running average window. The voltage measurement unit is driven by a phase-locked loop (PLL) to take into account variations of system frequency.

2) Voltage regulator uses a PI regulator to regulate primary voltage at the reference voltage (1.0 pu specified in the SVC Controller block menu). A voltage droop is incorporated in the voltage regulation to obtain a V-I characteristic with a slope (0.01 pu/100 MVA in this case). Therefore, when the SVC operating point changes from fully capacitive (+300 Mvar) to fully inductive (-100 Mvar) the SVC voltage varies between 1-0.03=0.97 pu and 1+0.01=1.01 pu..

3) Distribution Unit uses the primary susceptance Bsvc computed by the voltage regulator to determine the TCR firing angle α and the status (on/off) of the three TSC branches. The firing angle α as a function of the TCR susceptance BTCR is implemented by a look-up table from the equation

$$B_{TCR} = \frac{2(\pi-\alpha)+\sin(2\alpha)}{\pi} \quad (1)$$

where BTCR is the TCR susceptance in pu of rated TCR reactive power (109 Mvar).

4) Firing Unit consists of three independent subsystems, one for each phase (AB, BC and CA). Each subsystem consists of a PLL synchronized on line-to-line secondary voltage and a pulse

generator for each of the TCR and TSC branches. The pulse generator uses the firing angle α and the TSC status coming from the Distribution Unit to generate pulses. The firing of TSC branches can be synchronized (one pulse is sent at positive and negative thyristors at every cycle) or continuous.

The steady-state waveforms and the SVC dynamic response when the system voltage is varied is simulated in Fig. 3. Initially the source voltage is set at 1.004 pu, resulting in a 1.0 pu voltage at SVC terminals when the SVC is out of service. As the reference voltage Vref is set to 1.0 pu, the SVC is initially floating (zero current). This operating point is obtained with TSC1 in service and TCR almost at full conduction (α = 96 degrees).

At t=0.1s voltage is suddenly increased to 1.025 pu. The SVC reacts by absorbing reactive power (Q=-95 Mvar) to bring the voltage back to 1.01 pu. The 95% settling time is approximately 135 ms. At this point all TSCs are out of service and the TCR is almost at full conduction (α = 94 degrees).

At t=0.4 s the source voltage is suddenly lowered to 0.93 pu. The SVC reacts by generating 256 Mvar of reactive power, thus increasing the voltage to 0.974 pu. At this point the three TSCs are in service and the TCR absorbs approximately 40% of its nominal reactive power (α =120 degrees). It is observed how the TSCs are sequentially switched on and off. Each time a TSC is switched on the TCR α angle changes from 180 degrees (no conduction) to 90 degrees (full conduction).

Finally, at t=0.7 s the voltage is increased to 1.0 pu and the SVC reactive power is reduced to zero.

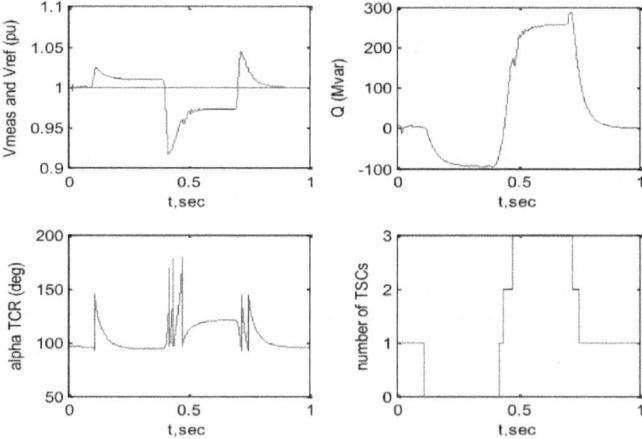

Fig. 3. Wafeforms illustrating SVC Dynamic response to System Voltage Steps

4. TRANSIENT STABILITY OF A POWER SYSTEM

This case study analyzes transient stability of a two-machine transmission system with Power System Stabilizers (PSS) and Static Var Compensator (SVC). Fig. 4 shows a 1000 MW hydraulic generation plant (machine M1) is connected to a load center through a long 500 kV, 700 km transmission line. The load center is modeled by a 5000 MW resistive load. The load is fed by the remote 1000 MW plant and a local generation of 5000 MW (machine M2). The system has been initialized so that the line carries 950 MW which is close to its surge impedance loading (SIL = 977 MW). In order to maintain system stability after faults, the transmission line is shunt compensated at its center by a 200-Mvar Static Var Compensator (SVC).

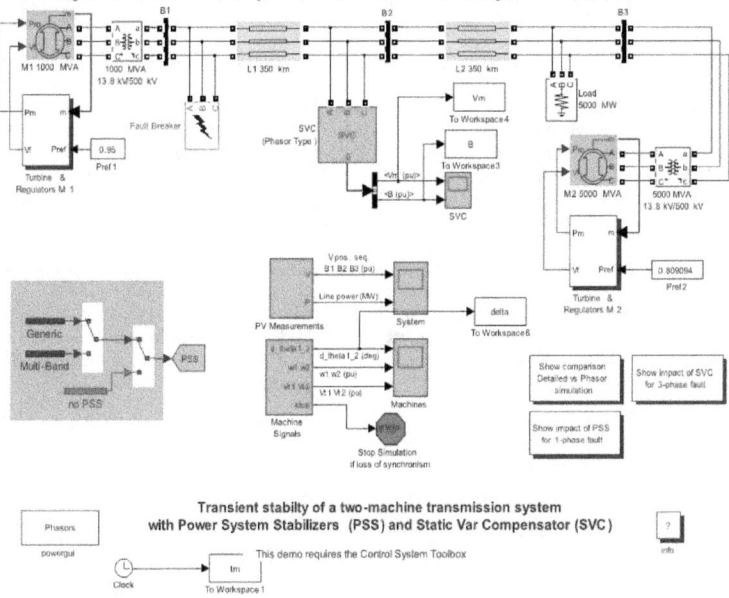

Fig. 4. Model of Transmission System

The SVC does not have a Power Oscillation Damping (POD) unit. The two machines are equipped with a Hydraulic Turbine and Governor (HTG), Excitation system and Power System Stabilizer (PSS). These blocks are located in the two 'Turbine and Regulator' subsystems. Two types of stabilizers can be selected: a generic model using the acceleration power (Pa= difference between mechanical power Pm and output electrical power Peo) and a Multi-band stabilizer using the speed deviation (dw). The stabilizer type can be selected by specifying a value (0=No

PSS 1=Pa PSS or 2= dw MB PSS) in the PSS constant block. The SVC is the phasor model from the FACTS library. In the Control parameters of SVC dialog, you can select either Voltage regulation or Var control (Fixed susceptance Bref) mode. Initially the SVC is set in Var control mode with a susceptance Bref=0, which is equivalent to having the SVC out of service. A Fault Breaker block is connected at bus B1. You will use it to program different types of faults on the 500 kV system and observe the impact of the PSS and SVC on system stability.

To start the simulation in steady-state, the machines and the regulators have been previously initialized by means of the Load Flow and Machine Initialization utility of the Powergui block. Load flow has been performed with machine M1 defined as a PV generation bus (V=13800 V, P=950 MW) and machine M2 defined as a swing bus (V=13800 V, 0 degrees). After the load flow has been solved, the reference mechanical powers and reference voltages for the two machines have been automatically updated in the two constant blocks connected at the HTG and excitation system inputs: Pref1=0.95 pu (950 MW), Vref1=1.0 pu; Pref2=0.8091 pu (4046 MW), Vref2=1.0 pu.

A 3-phase fault is applied and observed the impact of the SVC for stabilizing the network during a severe contingency. The two PSS are set in service (value=1 in the PSS constant block). The 'Fault Breaker' block is programmed in order to apply a 3-phase-to-ground fault. The SVC is in fixed susceptance mode with Bref = 0. By looking at the rotor angle difference between the two machines d_theta1_2 (delta 1 – delta 2) signal, it should be observed that the two machines quickly fall out of synchronism after fault clearing. The simulation results are shown in Fig. 5 for 3 phase faults and no SVC.

Then, the SVC mode of operation is changed to 'Voltage regulation'. The SVC will now try to support the voltage by injecting reactive power on the line when the voltage is lower than the reference voltage (1.009 pu). The chosen SVC reference voltage corresponds to the bus voltage with the SVC out of service. In steady state the SVC will therefore be 'floating' and waiting for voltage compensation when voltage departs from its reference set point. After restarting the simulation, it is observed that the system is now stable with a 3-phase fault. The simulation results are shown in Fig. 6 for 3 phase faults and SVC in voltage regulation mode.

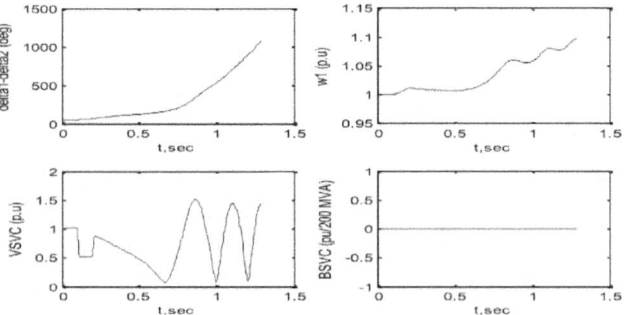

Fig. 5. Waveforms for 3 phase fautls and no SVC

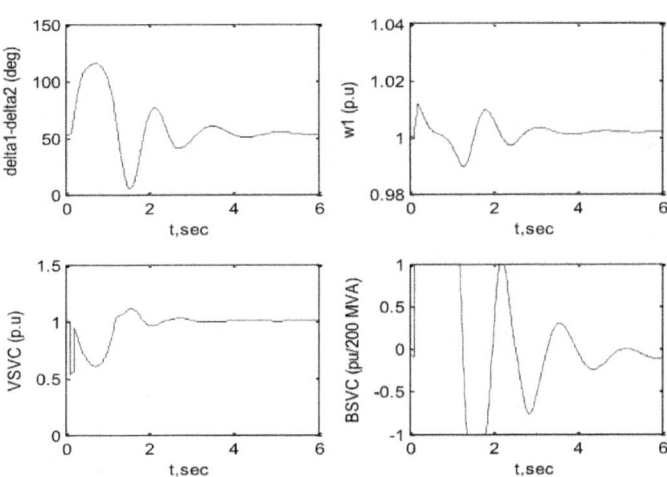

Fig. 6. Waveforms for 3 phase faults and SVC in voltage regulation mode

5. GTO-BASED STATCOM

The STATCOM consists of a three-level 48-pulse inverter and two series-connected 3000 µF capacitors which act as a variable DC voltage source. The variable amplitude 60 Hz voltage produced by the inverter is synthesized from the variable DC voltage which varies around 19.3 kV. Fig. 7 shows SPS Model of the 100 Mvar STATCOM on 500 kV Power System. Fig. 8 shows the shunt controller (48-pulses, 3 level inverters)

The control system task is to increase or decrease the capacitor DC voltage, so that the generated AC voltage has the correct amplitude for the required reactive power. The control system must also keep the AC generated voltage in phase with the system voltage at the STATCOM connection bus to generate or absorb reactive power only (except for small active power required by transformer and inverter losses).

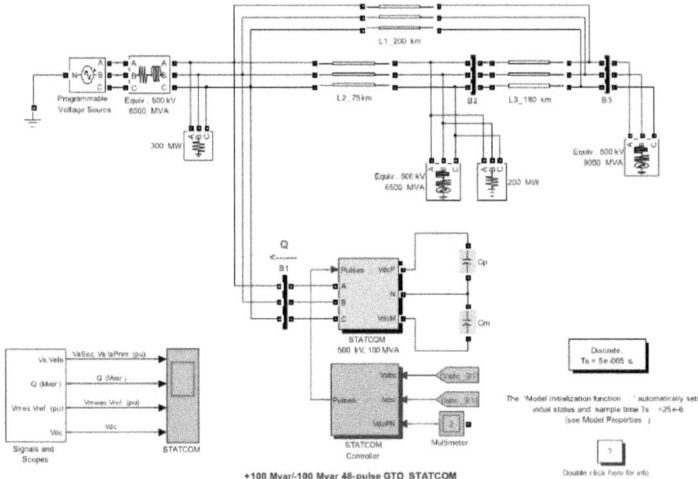

Fig. 7. SPS Model of the 100 Mvar STATCOM on 500 kV Power System

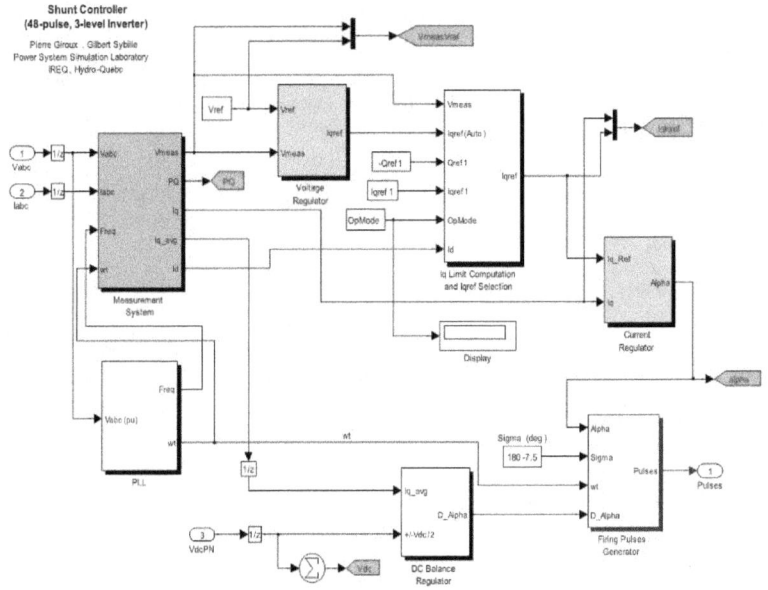

Fig. 8. Shunt Controller (48-pulses, 3 level inverters)

The control system uses the following modules

1) PLL (phase locked loop) synchronizes GTO pulses to the system voltage and provides a reference angle to the measurement system.
2) Measurement System computes the positive-sequence components of the STATCOM voltage and current, using phase-to-dq transformation and a running-window averaging.
3) Voltage regulation is performed by two PI regulators: from the measured voltage Vmeas and the reference voltage Vref, the Voltage Regulator block (outer loop) computes the reactive current reference Iqref used by the Current Regulator block (inner loop). The output of the current regulator is the α angle which is the phase shift of the inverter voltage with respect to the system voltage. This angle stays very close to zero except during short periods of time, as explained below.
4) A voltage droop is incorporated in the voltage regulation to obtain a V-I characteristics with a slope (0.03 pu/100 MVA in this case). Therefore, when the STATCOM operating point changes from fully capacitive (+100 Mvar) to fully inductive (-100 Mvar) the SVC voltage varies between 1-0.03=0.97 pu and 1+0.03=1.03 pu.
5) Firing Pulses Generator generates pulses for the four inverters from the PLL output (ω.t) and the current regulator output (α angle).

To explain the regulation principle, let us suppose that the system voltage Vmeas becomes lower than the reference voltage Vref. The voltage regulator will then ask for a higher reactive current output (positive Iq= capacitive current). To generate more capacitive reactive power, the current regulator will then increase α phase lag of inverter voltage with respect to system voltage, so that an active power will temporarily flow from AC system to capacitors, thus increasing DC voltage and consequently generating a higher AC voltage.

STATCOM dynamic response to system voltage steps is shown in Fig. 9. Initially the programmable voltage source is set at 1.0491 pu, resulting in a 1.0 pu voltage at bus B1 when the STATCOM is out of service. As the reference voltage Vref is set to 1.0 pu, the STATCOM is initially floating (zero current). The DC voltage is 19.3 kV.

At t=0.1s, voltage is suddenly decreased by 4.5% (0.955 pu of nominal voltage). The STATCOM reacts by generating reactive power (Q=+70 Mvar) to keep voltage at 0.979 pu. The 95% settling time is approximately 47 ms. At this point the DC voltage has increased to 20.4 kV. Then, at t=0.2 s the source voltage is increased to 1.045 pu of its nominal value. The STATCOM reacts by changing its operating point from capacitive to inductive to keep voltage at 1.021 pu. At this point the STATCOM absorbs 72 Mvar and the DC voltage has been lowered to 18.2 kV. Finally, at t=0.3 s the source voltage in set back to its nominal value and the STATCOM operating point comes back to zero Mvar.

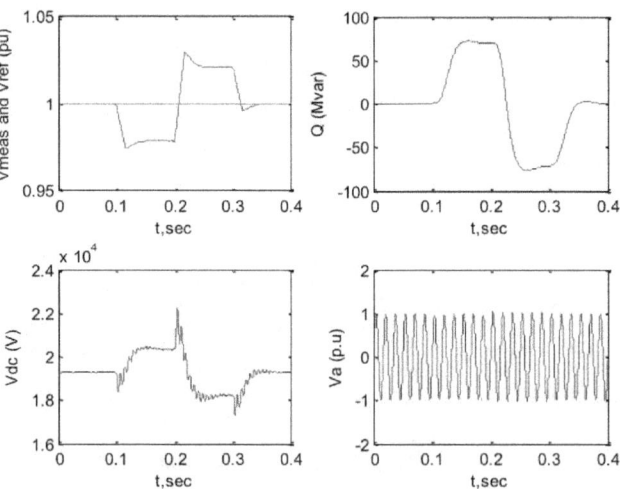

Fig. 9. Waveforms illustrating STATCOM dynamic response to system voltage steps

6. CONTROL OF LOAD FLOW USING UPFC

A UPFC that used to control the power flow in a 500 kV /230 kV transmission system is shown in Fig. 10. The system, connected in a loop configuration, consists essentially of five buses (B1 to B5) interconnected through transmission lines (L1, L2, L3) and two 500 kV/230 kV transformer banks Tr1 and Tr2. Two power plants located on the 230-kV system generate a total of 1500 MW which is transmitted to a 500-kV 15000-MVA equivalent and to a 200-MW load connected at bus B3. The plant models include a speed regulator, an excitation system as well as a power system stabilizer (PSS). In normal operation, most of the 1200-MW generation capacity of power plant #2 is exported to the 500-kV equivalent through three 400-MVA transformers connected between buses B4 and B5. For this case study we are considering a contingency case where only two transformers out of three are available (Tr2= 2*400 MVA = 800 MVA).

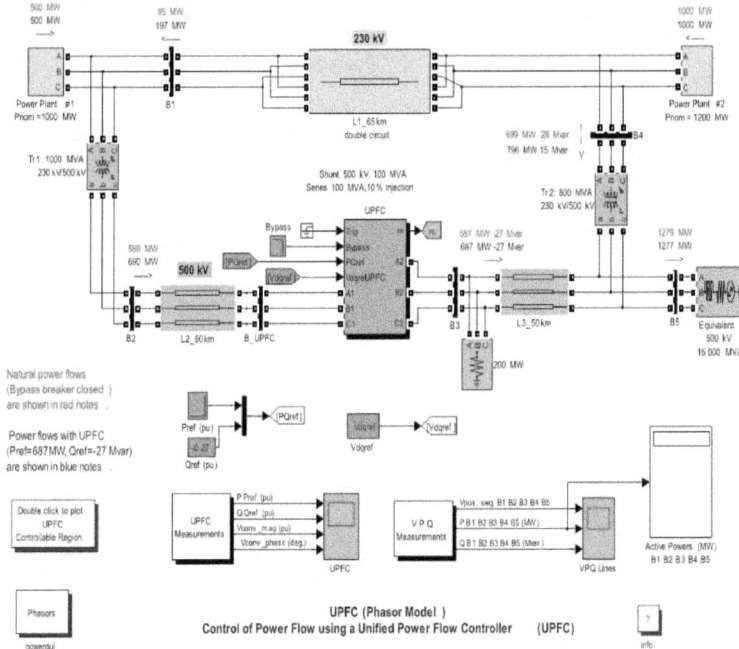

Fig. 10. Model of UPFC Controlling Power on a 500 KV / 230 KV Power System

Using the load flow option of the powergui block, the model has been initialized with plants #1 and #2 generating respectively 500 MW and 1000 MW and the UPFC out of service (Bypass breaker closed). The load flow shows that most of the power generated by plant #2 is transmitted through the 800-MVA transformer bank (899 MW out of 1000 MW), the rest (101 MW), circulating in the loop. Transformer Tr2 is therefore overloaded by 99 MVA. The demonstration illustrates how the UPFC can relieve this power congestion. The UPFC located at the right end of line L2 is used to control the active and reactive powers at the 500-kV bus B3, as well as the voltage at bus B_UPFC. It consists of a phasor model of two 100-MVA, IGBT-based, converters (one connected in shunt and one connected in series and both interconnected through a DC bus on the DC side and to the AC power system, through coupling reactors and transformers). Parameters of the UPFC power components are given in the dialog box. The series converter can inject a maximum of 10% of nominal line-to-ground voltage (28.87 kV) in series with line L2. The control parameters of the series converter are set "Mode of operation = Power flow control".

The UPFC reference active and reactive powers are set in the blocks labeled "Pref(pu)" and "Qref(pu)". Initially the Bypass breaker is closed and the resulting natural power flow at bus B3 is 587 MW and -27 Mvar. The Pref block is programmed with an initial active power of 5.87 pu corresponding to the natural flow. Then, at t=10s, Pref is increased by 1 pu (100 MW), from 5.87 pu to 6.87 pu, while Qref is kept constant at -0.27 pu.

Fig. 11 shows the UPFC system dynamic response to a change in reference power from 587 MW to 687 MW. P and Q measured at bus B3 follow the reference values. At t=5 s, when the Bypass breaker is opened the natural power is diverted from the Bypass breaker to the UPFC series branch without noticeable transient. At t=10 s, the power increases at a rate of 1 pu/s. It takes one second for the power to increase to 687 MW. This 100 MW increase of active power at bus B3 is achieved by injecting a series voltage of 0.089 pu with an angle of 94 degrees. This results in an approximate 100 MW decrease in the active power flowing through Tr2 (from 899 MW to 796 MW), which now carries an acceptable load.

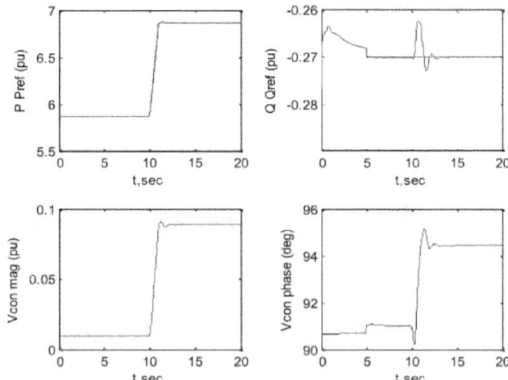

Fig. 11. UPFC dynamic response to a change in reference power from 587 MW to 687 MW

7. CONTROL OF AC MOTOR

The induction motor is fed by a current-controlled PWM inverter which is built using a Universal Bridge block is shown in Fig. 12. The motor drives a mechanical load characterized by inertia J, friction coeficient B, and load torque TL. The speed control loop uses a proportional-integral controller to produce the quadrature-axis current reference iq* which controls the motor torque. The motor flux is controlled by the direct-axis current reference id*. The induction motor is modeled by an Asynchronous Machine block. The motor used in this case study is a 50 HP, 460 V, four-pole, 60 Hz motor. Initially the reference speed is set a constant value of 120 rad/s and the load torque is also maintained constant at 0 N.m. The induction motor is fed by a current-controlled PWM inverter which is built using a Universal Bridge block. The IGBT inverter is modeled by a Universal Bridge block. The DC link input voltage is represented by a 780 V DC voltage source. The vector control for the inverter is shown in Fig. 13. The current regulator, which consists of three hysteresis controllers, is built with Simulink blocks. The motor currents are provided by the measurement output of the Asynchronous Machine block. The conversions between abc and dq reference frames are executed by the abc_to_dq0 Transformation and dq0_to_abc The rotor flux is calculated by the Flux_Calculation block. The rotor flux position (θ_e) is calculated by the Teta Calculation block. The stator quadrature-axis current reference (iqs*) is calculated by the iqs*_Calculation block. The stator direct-axis current reference (ids*) is calculated by the id*_Calculation. The speed controller is of proportional-integral type and is implemented using Simulink blocks.

Fig. 14. shows the motor voltage, current and torque waveforms obtained when the motor is running at no load (torque=0 N.m) at a speed of 120 rad / sec.

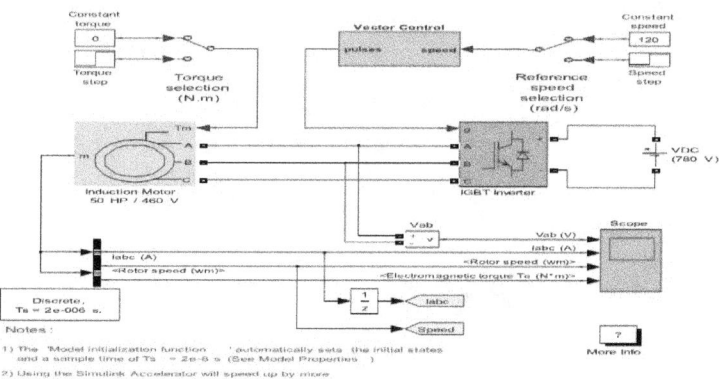

Fig. 12. Variable-Speed Field-Oriented Induction Motor

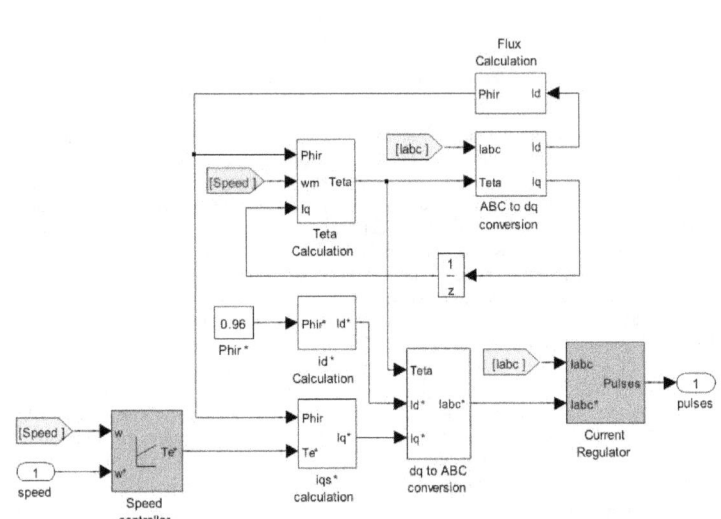

Fig. 13. Vector Control

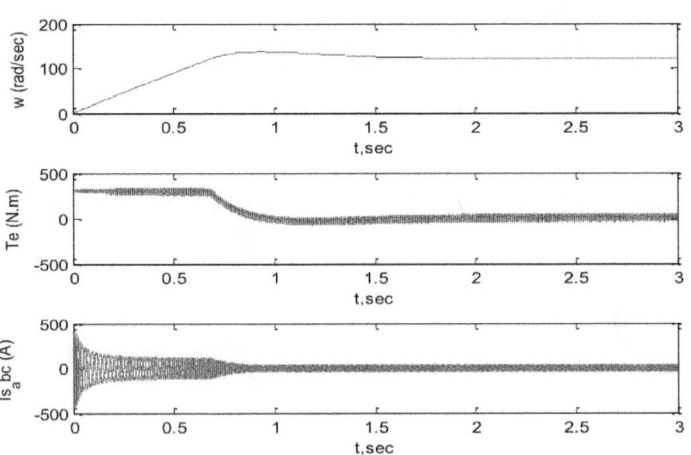

Fig. 14. Motor speed, torque and current waveforms

8. CONTROL OF DC MOTOR

Description of the drive system is shown in Fig. 15. The DC motor is fed by the DC source through a chopper that consists of the GTO thyristor, Th1, and the free-wheeling diode D1. The DC motor drives a mechanical load that is characterized by the inertia J, friction coefficient B, and load torque TL (which can be a function of the motor speed). Thyristor Th1 is triggered by a pulse-width-modulated (PWM) signal to control the average motor voltage. The motor torque is controlled by the armature current Ia, which is regulated by a current control loop. The motor speed is controlled by an external loop, which provides the current reference Ia* for the current control.

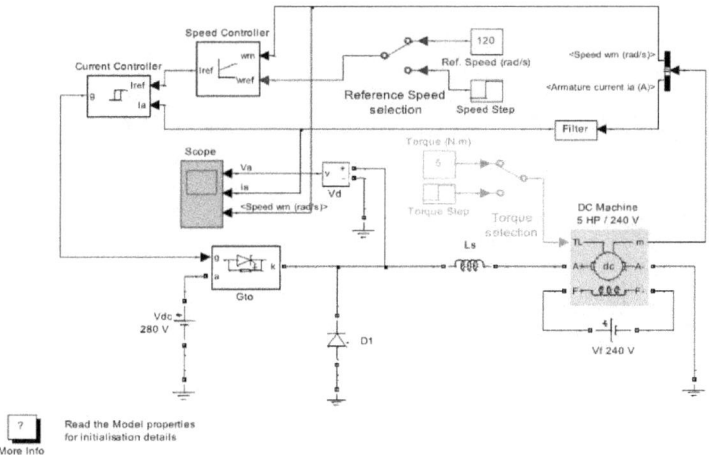

Fig. 15. Chopped-Fed DC Motor Drive

Initially the reference speed is set to a constant value of 120 rad/s and the load torque is also maintained constant at 5 N.m. The motor used in this case study is a separately excited, 5 HP/240 V DC motor. The required trigger signal for the GTO thyristor is generated by a hysteresis current controller, which forces the motor current to follow the reference. The starting transient of the DC drive is simulated. The speed reference is 120 rad/s, and you can observe the DC motor speed and current within +h/2 and -h/2 limits (h is the hysteresis band). The speed control loop uses a proportional-integral controller, which is implemented by Simulink blocks. The motor voltage, current waveforms, and motor speed are displayed in Fig. 16

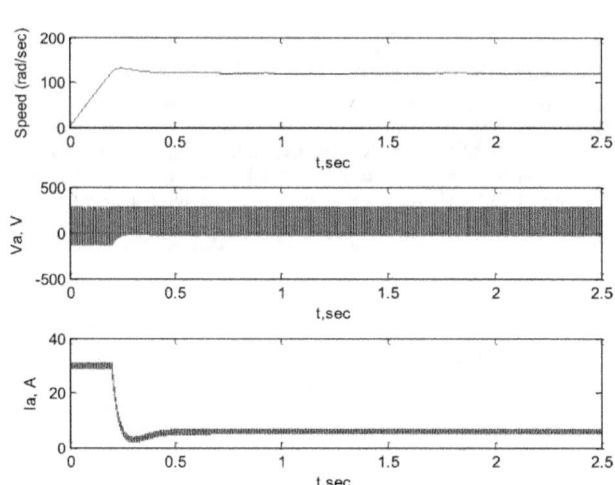

Fig. 16. Motor speed, voltage and current waveforms

9. VSC-Based HVDC Link

The principal characteristic of VSC-HVDC transmission is its ability to independently control the reactive and real power flow at each of the AC systems to which it is connected, at the Point of Common Coupling (PCC). In contrast to line-commutated HVDC transmission, the polarity of the DC link voltage remains the same with the DC current being reversed to change the direction of power flow. Fig. 17 shows the VSC-HVDC Transmission system Model. The rectifier and the inverter are interconnected through a 75 km cable (2 pi sections). A circuit breaker is used to apply a three-phase to ground fault on the inverter AC side.

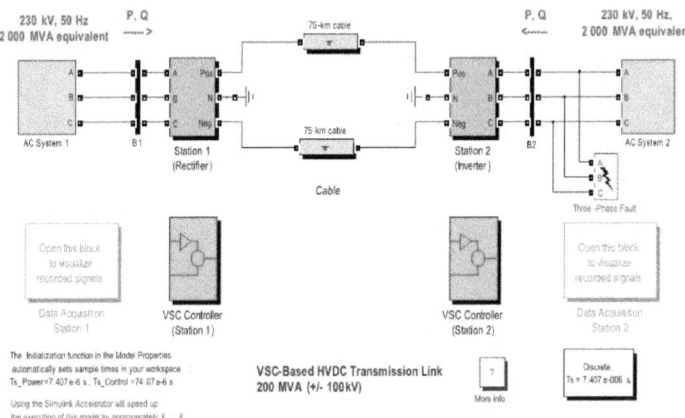

Fig. 17. VSC-HVDC Transmission system Model

The two controllers are independent with no communication between them. Each converter has two degrees of freedom. In our case, these are used to control:
-P and Q in station 1 (rectifier)
-Udc and Q in station 2 (inverter).

The converter 1 and converter 2 controllers design are identical and is shown in Fig. 19. The Phase Locked Loop block measures the system frequency and provides the phase synchronous angle θ. The active and reactive power and voltage loop contains the outer loop regulators that calculate the reference value of the converter current vector (Iref_dq) which is the input to the inner current loop. The control modes are: in the "d" axis, either the active power flow at the PCC or the pole-to-pole DC voltage; in the "q" axis, the reactive power flow at the PCC. Note that, it would be also possible to add an AC voltage control mode at the PCC in the "q" axis. The main functions of Inner Current Loop block, the AC Current Control block tracks the current reference vector ("d" and "q" components) with a feed forward scheme to achieve a fast control of the current at load changes and disturbances. In essence, it consist of knowing the U_dq vector voltages and computing what the converter voltages have to be, by adding the voltage drops due to the currents across the impedance between the U and the PWM-VSC voltages. The state equations representing the dynamics of the VSC currents are used. The "d" and "q" components are decoupled to obtain two independent first-order plant models. A proportional integral (PI) feedback of the converter current is used to reduce the error to zero in steady state. The output of the AC Current Control block is the unlimited reference voltage vector Vref_dq_tmp. The DC Voltage Balance Control can be enabled or disabled. The difference between the DC side voltages are controlled to keep the DC side of the three level bridge balanced in steady-state. The DC midpoint current Id_0 determines the difference Ud_0 between the upper and lower DC voltages. The Reference Voltage Conditioning block takes into account the actual DC voltage and the theoretical maximum peak value of the fundamental bridge phase voltage in relation to the DC voltage to generate the new optimized reference voltage vector. The Reference Voltage Limitation block limits the reference voltage vector amplitude to 1.0, since over modulation is not desired. The Inverse dq and Inverse Clark transformation blocks are required to generate the three-phase voltage references to the PWM.

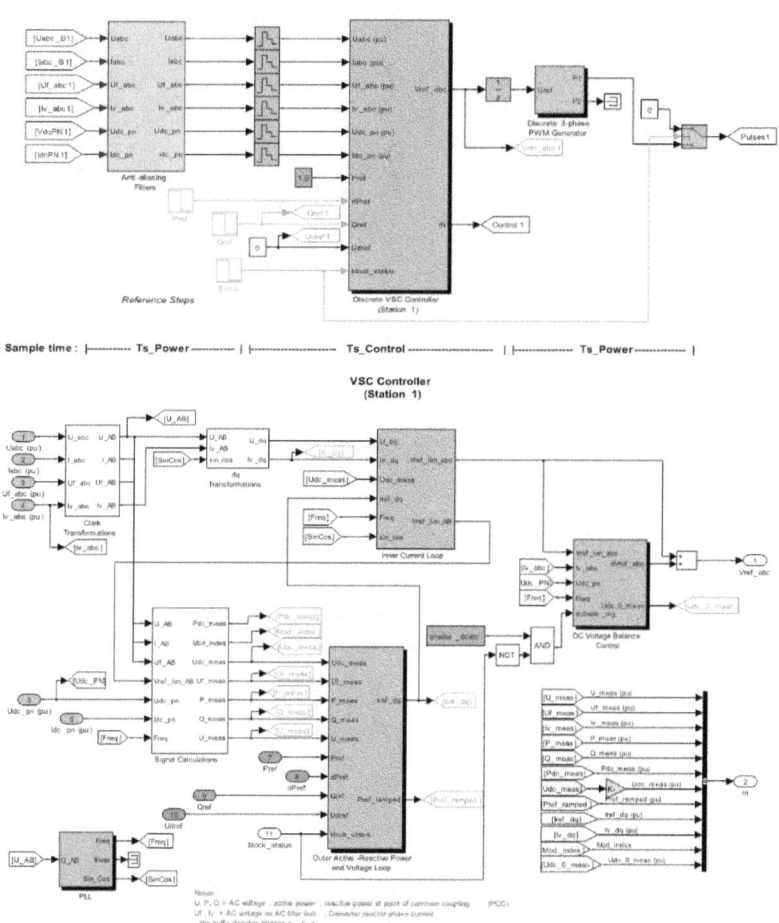

Fig. 19. Station I VSC Controller

Two simulations will permit to examine the system response to was carried on.

1) Steps on the regulators references: The system is programmed to start and reach a steady state. Steps are then applied sequentially on: the reference active and reactive power of the rectifier; the reference DC voltage of the inverter. The dynamic response of the regulators is observed and shown in Fig. 20. At t = 1.5 s, a 0.1 p.u. step is first applied to the reference active power (decrease from 1 p.u. to 0.9 pu). The power stabilizes in approximately 0.3 seconds. Steps are also applied to the reference reactive power of the rectifier (from 0 to -0.1 p.u.) at t = 2.0 s and on the reference DC voltage of the inverter (decrease from 1 p.u. to 0.95 p.u.) at t = 2.5 s. Note the regulators dynamics and how they are more or less mutually affected. The control design attempts to decouple the active and reactive power responses.

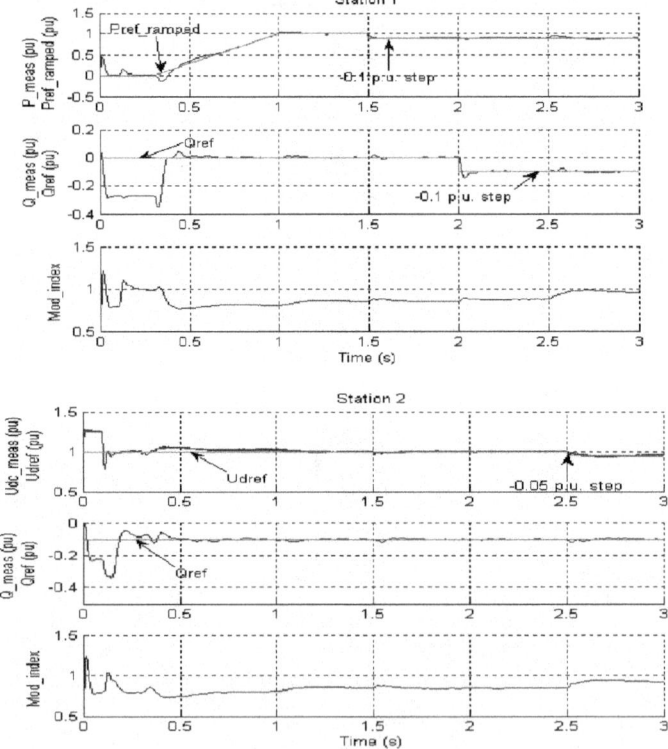

Fig. 20. Startup and P & Q & Udc responses of station 1 & 2

2) Minor and severe perturbations on the AC sides: Voltage sag was programmed so the source will have a step change of -0.1 p.u on voltage magnitude at t = 1.5 s for a duration of 7 cycles. The system dynamics responses are shown in Fig. 21. After the AC voltage sag in station 1, the active and reactive power deviation from the pre-disturbance is less than 0.09 p.u. and 0.2 p.u. respectively. The recovery time is less than 0.3 s and steady state is reached again. A second perturbation follows.

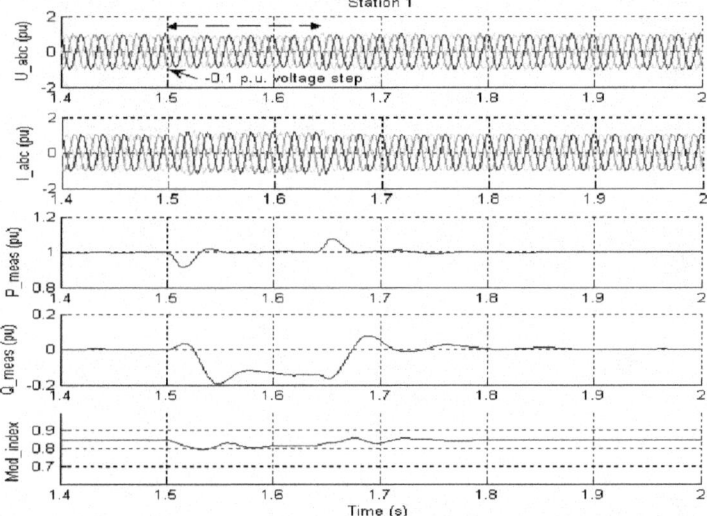

Fig. 21. Voltage step on system 1

A 6 cycles three-phase fault will be applied at t = 2.1 s in station 2 PCC (Bus B2). The system dynamics responses are shown in Fig. 22. During the severe three-phase fault at station 2, the transmitted DC power is almost halted and the DC voltage tends to increase (1.2 p.u.) since the DC side capacitance is being excessively charged. A special function (DC Voltage Control Override) in the Active Power Control (in station 1) attempts to limit the DC voltage within a fixed range. The system recovers well after the fault within 0.5 s. You can observe overshoot in the active power (1.33 p.u. in station 1) and damped oscillations (around 10 Hz) in the reactive power.

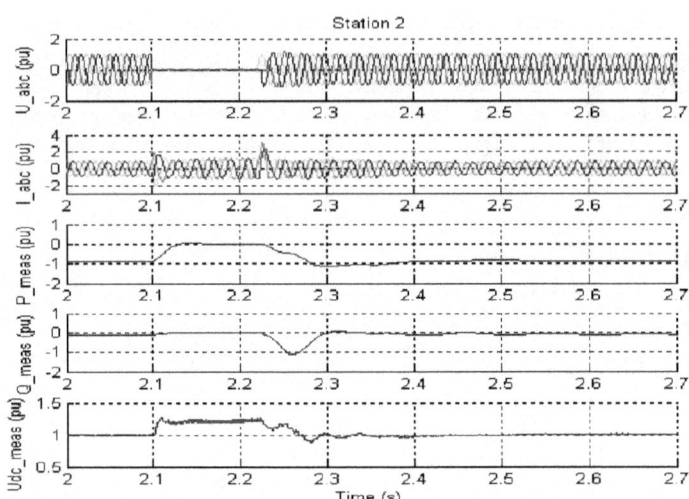

Fig. 22. Three Phase to Ground at station 2 Bus

10. CONCLUSION

SimPowerSystems is a modern design tool that allows scientists and engineers to rapidly and easily build models that simulate power systems The paper covered some case studies that provide detailed, realistic examples of how to use SimPowerSystems in the simulation of various power system applications with power electronic devices. The following case studies were presented in the paper:
1-Thyristor-Based Static Var Compensator.
2. Transient Stability of a Power System with SVC and PSS.
3. GTO-Based STATCOM.
4. Control of load flow using UPFC.
5- Control of AC motor.
6- Control of DC motor.
7- VSC-Based HVDC Link

It was noted is that the SimPowerSystems software can be good tool in simulating the dynamics of small power systems, but as the system become large there will be a lot of complications. The SimPowerSystems software does not provide other types of static analysis for power system that are available with other types of simulators such as PowerWorld and SimPow simulators, i.e., power flow study, with economic dispatch and optimal load flow AGC modes, fault analysis and contengency analysis. So, the choice of the simulator will depend on the type of study that is needed.

11. REFERENCES

[1] The Mathworks (www.mathworks.com), SimPowerSystems documentation Version V5.2 (R2009b)
[2] Narian G. Hingorani, Laszlo Gygugyi, "Understanding FACTS"
[3] Anderson, Fouad, A. A., "Power system control and stability" , 1977
[4] Samir K. datta, Reston Pub. Co, "Power Electronics and Control", 1985,
[5] Dr. Hedaya alasooly, "PhD thesis: Modeling and control of FACTS devices for power system quality improvement", Electrical Energy Department, Czech Technical University, 2003, published in "www.lulu.com

www.ingramcontent.com/pod-product-compliance
Lightning Source LLC
Chambersburg PA
CBHW051533240526
45471CB00019B/1323